Volcanic and Seismic Hazards on the Island of Hawaii

Lava flows entered Kalapana Gardens in December 1986.

Front Cover: View of Kapoho village during the 1960 eruption before it was entirely destroyed. *(Photographer unknown)*

Inside Front Cover *(Photograph by J.D. Griggs)*

For sale by the U.S. Government Printing Office
Superintendent of Documents, Mail Stop: SSOP, Washington, DC 20402-9328
ISBN 0-16-038200-9

Preface

The eruptions of volcanoes often have direct, dramatic effects on the lives of people and on their property. People who live on or near active volcanoes can benefit greatly from clear, scientific information about the volcanic and seismic hazards of the area. This booklet provides such information for the residents of Hawaii so they may effectively deal with the special geologic hazards of the island.

Identifying and evaluating possible geologic hazards is one of the principal roles of the U.S. Geological Survey (USGS) and its Hawaiian Volcano Observatory. When USGS scientists recognize a potential hazard, such as an impending eruption, they notify the appropriate government officials, who in turn are responsible for advising the public to evacuate certain areas or to take other actions to insure their safety.

This booklet was prepared in cooperation with the Hawaii County Civil Defense Agency.

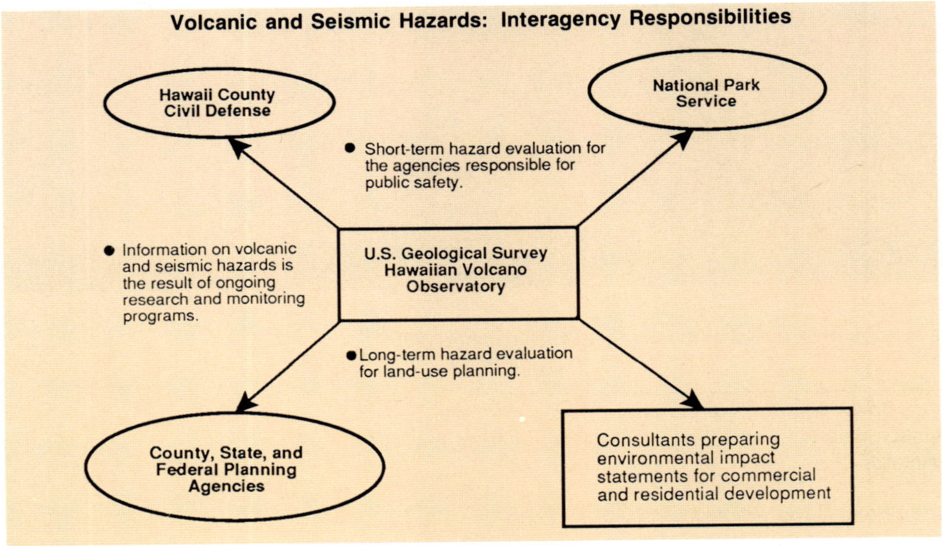

Introduction

The Hawaiian Islands are at the southeastern end of a chain of volcanoes that began to form over 70 million years ago. Each island is made up of one or more volcanoes that first erupted at the bottom of the sea and emerged above the ocean's surface after countless eruptions over hundreds of thousands of years. Most of the volcanic activity in the last two centuries has occurred on the Island of Hawaii, the youngest island of the chain. Hawaii is composed of five volcanoes, at least two of which–Kilauea and Mauna Loa–are expected to erupt frequently throughout the foreseeable future.

The Hawaiian Islands are the tops of volcanoes that rise as much as 30,000 feet above the sea floor. *(Map used by permission of Dynamic Graphics, Berkeley, California.)*

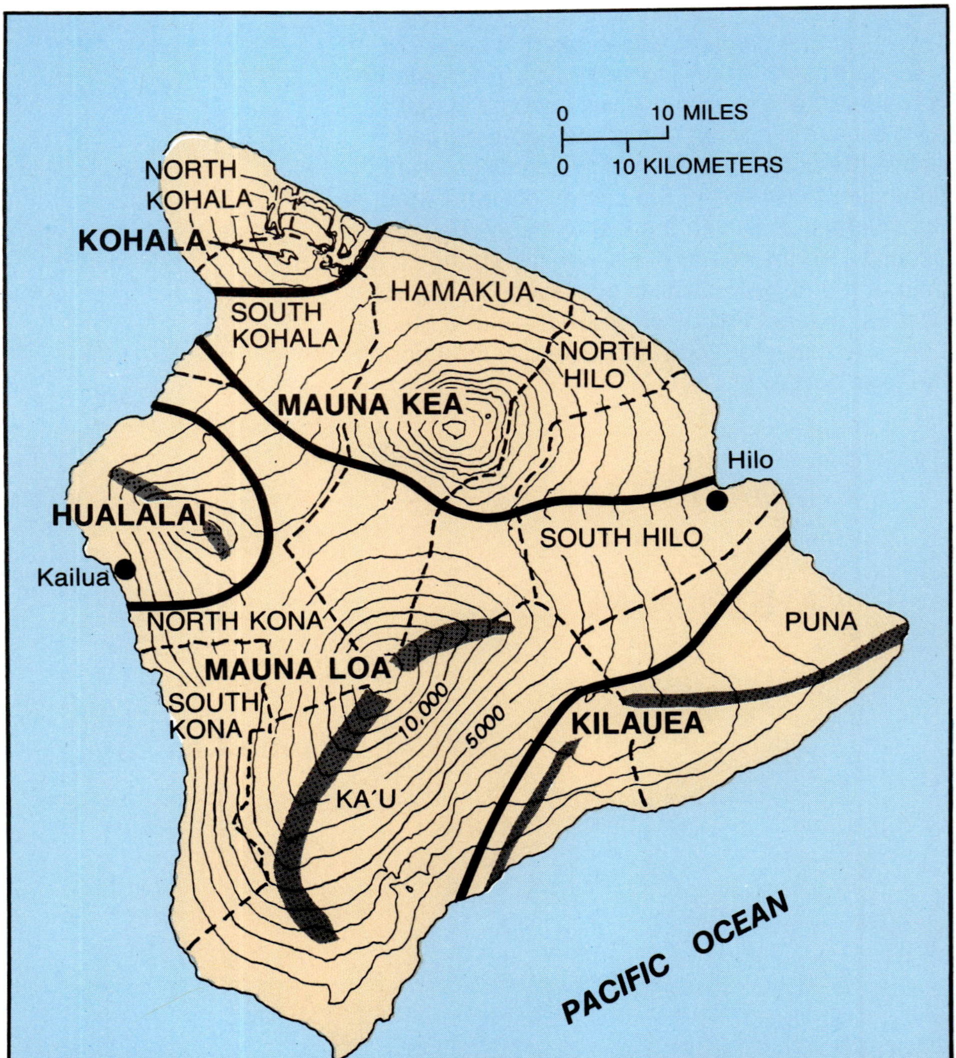

The Island of Hawaii and its five volcanoes. The rift zones of the historically active volcanoes are indicated by the stippled gray pattern. Dashed lines indicate the boundaries of districts on the island.

The Island of Hawaii owes its unique landscape to its volcanic origin. From the barren, red cinder cones on the summit of Mauna Kea to the fern-lined craters and black sand beaches of Puna, volcanic activity has combined with the subtropical climate to create the scenery that draws visitors from all over the world. Hawaii's residents not only enjoy the benefits of past volcanic activity, but also have the opportunity to witness the frequent eruptions of the active volcanoes as the island continues to grow.

Living on the slopes of an active volcano, however, brings its own special hazards. As more of Hawaii's land area is developed, the likelihood of volcanic eruptions interfering with human activity increases. The direct hazards produced by volcanic eruptions include lava flows, airborne lava fragments, and volcanic gases. Indirect hazards, such as earthquakes and tsunamis (earthquake-generated sea waves), may cause damage over an even greater area.

Information in this booklet on hazards on the Island of Hawaii will help residents to make decisions about where to buy property, build a house, or start a business. A family planning to build a home may decide that the advantages of an area, such as the beauty of a beach lot, outweigh the risks, even in a high hazard zone. But such decisions can be made intelligently only if the hazards are understood beforehand.

Public officials who must make decisions regarding land-use planning will also find the information in this publication useful. The Hawaii County government receives numerous proposals for development projects ranging from resort complexes to space launch facilities. Obviously, the degree of volcanic and seismic hazard at a given site should be a major factor in evaluating these projects.

Volcanic Hazards

Hawaiian volcanoes can erupt either at their summits or on their flanks. Young Hawaiian volcanoes, such as Kilauea and Mauna Loa, have summit calderas. A caldera is a crater several miles in diameter that forms as the result of collapse when magma drains from beneath the summit. (*Magma* is the term used for molten rock that is still beneath the earth's surface; it is called *lava* when it reaches the surface.) Summit eruptions of Kilauea and Mauna Loa occur within or near their calderas. Flank eruptions usually take place along rift zones, which are highly fractured zones of weakness within the volcano. Rift zones typically extend from the summit of a volcano toward the coastline and may continue for many miles under the sea.

The Pu'u 'O'o vent began erupting in 1983 on Kilauea's east rift zone, 12 miles from the volcano's summit. *(Photograph provided by the National Park Service)*

Lava flows from the Kupaianaha vent on Kilauea's east rift zone bury a street intersection in Kalapana Gardens subdivision in December 1986. *(Photograph by J.D. Griggs, USGS)*

Lava flows are the most common of the direct hazards created by Hawaiian eruptions and pose the greatest threat to property. Other hazards include airborne particles of ash, cinder, and fragile strands of volcanic glass called Pele's hair, and corrosive volcanic gases. Explosive eruptions occur much less often than non-explosive eruptions at Hawaiian volcanoes but have been witnessed in historic time. The greatest danger associated with explosive eruptions is their potential to produce pyroclastic surges. These surges are highly destructive, turbulent gas clouds that flow rapidly along the ground carrying hot ash and rock fragments. A lesser volcanic hazard is created by ground movement, which may result in large cracks across roads and other property or cause uneven settling of foundations. Generally, only areas near an active or recently active volcanic vent are affected by large-scale ground cracks and settling.

Lava flows

Lava flows are the most common of the direct volcanic hazards in Hawaii. Flows may endanger people's property, livelihood, and peace of mind, but seldom their lives. The fronts of Hawaiian lava flows generally move more slowly than the speed at which people walk, although the lava in the channel behind the front may be flowing much faster. On steep slopes a large flow could travel rapidly enough to endanger persons in its path. During the 1950 eruption of Mauna Loa, a flow front advanced at an average speed of almost 6 mph for over 2 hours.

The speed of a lava flow is determined not only by the steepness of the terrain, but also by the volume of lava that is erupted, because large flows tend to advance more rapidly than do small flows. The distance that a flow travels ultimately depends both on the eruption rate and on the duration of the eruption.

The chemical composition of lava will also affect how rapidly a flow travels. Most Hawaiian lavas are classified as basalts, but within this

A lava flow from the Pu'u 'O'o vent advances down a street in the Royal Gardens subdivision in April 1983. *(Photograph by J.D. Griggs, USGS)*

Lava flows from Pu'u 'O'ō had already cut a swath through the Royal Gardens subdivision when this photograph was taken in 1983. Many more flows were to follow. The subdivision was developed on a steep slope below Kilauea's east rift zone. *(Photograph by J.D. Griggs, USGS)*

category there are many types. Some basalts are more fluid and will flow at greater speeds than others. The eruption of Hualalai in 1800-1801 produced lava flows that appear to have been more fluid than flows from similar eruptions on Kilauea and Mauna Loa.

The Pu'u 'O'o-Kupaianaha eruption on Kilauea's east rift zone, which began in 1983, provides a good example of two common, but very different, types of eruptive behavior: rapidly-moving flows produced during brief, high-volume eruptions, and slow-moving flows created by a prolonged low-volume eruption. The episodic eruptions at the Pu'u 'O'o vent, which was active from June 1983 through June 1986, produced a large volume of lava within a few hours' time. These outbursts were characterized by spectacular lava fountains and lava flows that moved rapidly down the volcano's south flank. The flows entered the Royal Gardens subdivision during 7 episodes and destroyed 16 homes. Each flow was short-lived, however, and stagnated soon after the lava fountains died. None of these flows reached the coastline.

In July 1986, the site of the eruption shifted to the Kupaianaha vent, 1.8 miles to the northeast of Pu'u 'O'o. Kupaianaha erupted almost continuously for over 5 years but at a much lower rate than Pu'u 'O'o. During the first few months of activity at Kupaianaha, the lava flows did not advance more than a mile beyond the vent. But after months of continuous eruption, a lava tube system formed as channeled lava flows gradually formed roofs, enclosing the rivers of lava within. Lava tubes are significant in terms of hazard, since the tubes insulate the lava and allow it to flow much farther before cooling and stopping.

An active lava pond overlay the Kupaianaha vent. Lava entered the lava tube system at the far end of the long neck on the pond. *(Photograph by E.W. Wolfe, USGS)*

A Kalapana house is ignited by a lava flow. By 1991, this community lay buried beneath 50-75 feet of lava. *(Photograph by J.B. Stokes, USGS)*

The hazards posed by a prolonged low-volume eruption soon became apparent as lava tubes from Kupaianaha extended toward the Kalapana coast. From November 1986 to October 1991, tube-fed flows repeatedly engulfed residential areas on the coastal plain, destroying 165 houses. Although these flows buried many acres within a single day, there was ample time to evacuate residents. Warnings issued by the Hawaii County Civil Defense allowed people enough time to remove most of their belongings and, in some cases, even to dismantle and move their homes. In 1992, the threat to inhabited areas eased when the eruption shifted to new vents on the southwest flank of the Pu'u 'O'o cone, inside Hawaii Volcanoes National Park.

The chief threat of lava flows to property owners is that the flows may burn structures and bury land. Even houses that are spared by the lava, however, may be rendered uninhabitable when the roads and utility lines leading to them are destroyed. By 1997, lava flows from Kilauea's eruption had covered 8 miles of the coastal highway, isolat-ing the few structures that remained within the area.

Airborne lava fragments

Most volcanic eruptions produce fragments of lava that are airborne for at least a short time before being deposited on the ground. These fragments are called "tephra," and include ash, cinders, and Pele's hair. In Hawaii, tephra is usually ejected by lava fountains and poses a serious hazard only in the immediate vicinity of an erupting

The high lava fountains of Pu'u 'O'o produced black clouds of tephra, cooled bits of lava that collected downwind of the vent to form deposits of cinders and Pele's hair.

A close-up view of tephra, including cinders and Pele's hair (lower left). *(Photograph by J.D. Griggs, USGS)*

vent. Windborne tephra, however, can be disruptive at greater distances. The combination of high lava fountains and strong winds may result in tephra being carried many miles downwind of the eruption site. During lava fountaining episodes at Pu'u 'O'o from 1984 to 1986, the prevailing trade winds deposited most of the tephra in remote areas of Hawaii Volcanoes National Park, but small particles reached the town of Naalehu 39 miles away. During the same episodes, Kona winds (from the southwest) occasionally carried tephra to Hilo, 22 miles from the vent.

The small amount of tephra that fell on inhabited areas was not harmful to most people, but it was a source of irritation to those with respiratory problems and an inconvenience to the many residents with rain-water-catchment systems. Following at least three high-fountaining episodes, Hawaii County Civil Defense recommended that people disconnect and clean their rain-water-catchment systems to prevent the particles from washing into their water supply.

Volcanic gases

Volcanic gases are emitted during all types of eruptions. Gases also can be released during repose periods by inactive eruptive vents and by fumaroles, vents that may never have produced any lava. The gas plume rising from an active vent on Kilauea consists of about 80 percent water vapor with lesser amounts of sulfur dioxide, carbon dioxide, and hydrogen. Small quantities (typically less than 1 percent by volume) of carbon monoxide, hydrogen sulfide, and hydrogen fluoride are also present. Extremely small amounts of mercury and other metals have been detected in gases emitted from vents along the east rift zone of Kilauea, but none have been found in concentrations large enough to create a direct health hazard.

Any hazard posed by volcanic gases is greatest immediately downwind from active vents; the concentration of the gases quickly diminishes as the gases mix with air and are carried by winds away

Each episode of high fountaining from Pu'u 'O'o released a tremendous plume of volcanic gas into the atmosphere. *(Photograph by J.D. Griggs, USGS)*

from the source. Brief exposure to gases near vents generally does not harm healthy people, but it can endanger those with heart and respiratory ailments, such as chronic asthma.

A common gas produced during Hawaiian eruptions that is potentially harmful to human health is sulfur dioxide. Even small concentrations of sulfur dioxide can combine with water to form sulfuric acid, which can attack skin, cloth, metal, and other materials. When a volcanic plume mixes with atmospheric moisture, acid rain results. Acid rain can significantly retard the growth of cultivated or natural plant life downwind of a vent that degasses over a long period of time.

The sulfur dioxide emitted from Kilauea's summit during typical non-eruptive periods affects a relatively small area downwind of the summit. Similarly, the gases produced during short-lived eruptions affect only a limited area, although their odor may be detected many miles from the vent. The continuous emission of volcanic fumes during Kilauea's Pu'u 'O'o-Kupaianaha eruption, however, resulted in persistent volcanic haze and acid-rain conditions in the South Kona district on the leeward side of the island.

In late 1987, studies conducted on private water-catchment systems in the South Kona area revealed higher than average acidity in several water samples. Drinking the acidic water does not pose a health hazard, but such water can leach lead from the lead roof flashings, lead-headed nails, and solder connections found in many plumbing systems, resulting in unsafe levels of lead in the drinking water. Extensive testing in 1988 determined that many water-catchment systems on the island, particularly those in the districts adjacent to or downwind of the active vent, contained elevated levels of lead. Residents with rain-catchment systems should contact the Hawaii State Department of Health for information on how to avoid lead contamination of their drinking water.

Broccoli plants damaged by volcanic fumes from the Pu'u 'O'o vent, 12 miles away. *(Photograph by J.D. Griggs, USGS)*

Volcanic fumes can also damage agricultural crops. During the 1969-74 eruption of Kilauea's Mauna Ulu vent, the South Kona district experienced prolonged periods of eruption-related smog. A study conducted in 1972 by the University of Hawaii's Agricultural Experiment Station at Hilo concluded that the acid rain resulting from the fume was responsible for severe damage to the Kona tomato crop. The Pu'u 'O'o-Kupaianaha eruption of Kilauea caused similar problems for vegetable and flower growers in both the Kona and Puna districts, who reported light-to-moderate crop damage during periods when winds blew the gases over their fields.

Explosive eruptions

The rare explosive eruptions in Hawaii generally are caused by the interaction of magma and ground water. The magnitude of the resulting steam explosion varies from harmless to catastrophic. Small steam-blast explosions occurred during the 1960 Kapoho eruption when the magma beneath the vents, which were near sea level, encountered saltwater trapped in the surrounding rocks. These steam blasts ejected black clouds of pulverized rock fragments but were of little hazard except to scientists working close to the vents.

A much larger steam-blast eruption occurred at the summit of Kilauea in 1924, when ground water apparently flowed into the heated rocks beneath the Halemaumau vent, which had been erupting nearly continuously for over a century. The explosions continued at intervals for 2 weeks, carpeting the area around Halemaumau crater with large rocks and a thin layer of ash. Boulders weighing several tons were thrown as far as 3,000

The 1924 explosive eruption of Kilauea mainly affected the immediate vicinity of Halemaumau crater. *(Photograph courtesy of the Bishop Museum)*

feet from the crater. The greatest hazard posed by this type of activity is that it may start abruptly and endanger unwary onlookers. The 1924 eruption claimed one fatality—a man who ventured too close to the vent between explosions to take photographs and was struck by a rock when the activity suddenly resumed.

The largest explosive eruption on Hawaii within historical time occurred in 1790. This eruption produced pyroclastic surges (turbulent clouds of hot gas and rock fragments) that originated at Kilauea's summit and flowed several miles to the southwest. Pyroclastic surges are extremely dangerous because they move at speeds of 30 to 200 mph, and humans and animals caught in their path are killed by either asphyxiation or heat. A band of Hawaiian warriors traveling from Hilo to the Ka'u district to battle with Chief Kamehameha were overtaken by one of the 1790 pyroclastic surges, and about 80 of them were killed. The 1790 eruption left deposits of rock fragments and ash up to 30 feet thick on the rim of Kilauea's summit caldera.

The thick deposits of ash exposed at many sites on the island indicate that even larger explosive eruptions occurred in prehistoric times and probably originated from Mauna Kea as well as from Kilauea. Explosive eruptions of any size take place infrequently in Hawaii, but the possibility of one occurring in our lifetime should not be totally discounted. Such eruptions are unlikely to begin without some warning. The most widespread hazard from an explosive eruption would be windborne ash, which could damage structures, machinery, and agricultural crops.

Ground cracks and settling

Ground cracks and settling are commonly associated with volcanic activity; both generally occur near active or recently active volcanic vents as the result of shallow underground movement of magma. The beginning of an eruption at a new site is preceded by cracking of the ground as magma is forcefully injected into the area. The cracks may be as much as 6 feet wide and over a mile long; typically they form within a period of hours. The Kapoho area on Kilauea's lower east rift zone experienced such ground breakage prior to eruptions in 1924, 1955, and 1960.

Before the 1960 eruption, a large ground crack opened in the village of Kapoho, which was built on Kilauea's east rift zone. The village was later completely destroyed by the eruption. *(Photograph provided by the National Park Service)*

Ground cracks associated with eruptive activity on Kilauea's upper southwest rift zone. *(Photograph by J.D. Griggs, USGS)*

Ground settling may occur near a vent at the end of an eruption as magma drains away from beneath the vent area. This process produces both small depressions and large collapse features, such as the pit craters and summit calderas of Kilauea and Mauna Loa. In either case, the subsidence may be gradual or abrupt.

The hazard presented by ground cracks and settling associated with eruptions is usually limited to areas near the active vent and thus is overshadowed by the hazard posed by lava flows. Man-made structures that escape other damage from an eruption, however, can be damaged or destroyed by cracking, tilting, or settling of the ground beneath them. Ground cracks will remain after the eruption is over and can pose a threat to unwary people and animals if the cracks are obscured by heavy vegetation.

Lava Flow Hazard Zone Maps

Maps showing volcanic hazard zones on the island of Hawaii were first prepared in 1974 by Donal Mullineaux and Donald Peterson of the U.S. Geological Survey and were revised in 1987 and 1992. The current map divides the island into zones that are ranked from 1 through 9 based on the probability of coverage by lava flows. Other direct hazards from eruptions, such as tephra fallout and ground cracking and settling, are not specifically considered on this map; however, these hazards also tend to be greatest in the areas of highest hazard from lava flows.

Hazard zones from lava flows are based chiefly on the location and frequency of both historic and prehistoric eruptions. "Historic eruptions" include those for which there are written records, beginning in the early 1800's, and those that are known from the oral traditions of the Hawaiians. Our knowledge of prehistoric eruptions is based on geologic mapping and dating of the old flows of each volcano. The hazard zones also take into account the larger topographic features of the volcanoes that will affect the distribution of lava flows. Finally, any hazard assessment is based on the assumption that future eruptions will be similar to those in the past.

Hazard zone boundaries are approximate. The change in the degree of hazard from one zone to the next is generally gradual rather than abrupt, and the change can occur over the distance of a mile or more. Within a single hazard zone, the severity of hazard may vary on a scale too fine to map. These variations may be the result of gradual changes that extend across the entire zone. For example, the hazard posed by lava flows decreases gradually as the distance from vents increases.

There may be abrupt changes, however, in the relative hazard because of the local topography. For example, the hills behind Ninole stand high above the adjacent slopes of Mauna Loa and consequently are at a much lower risk from lava flows

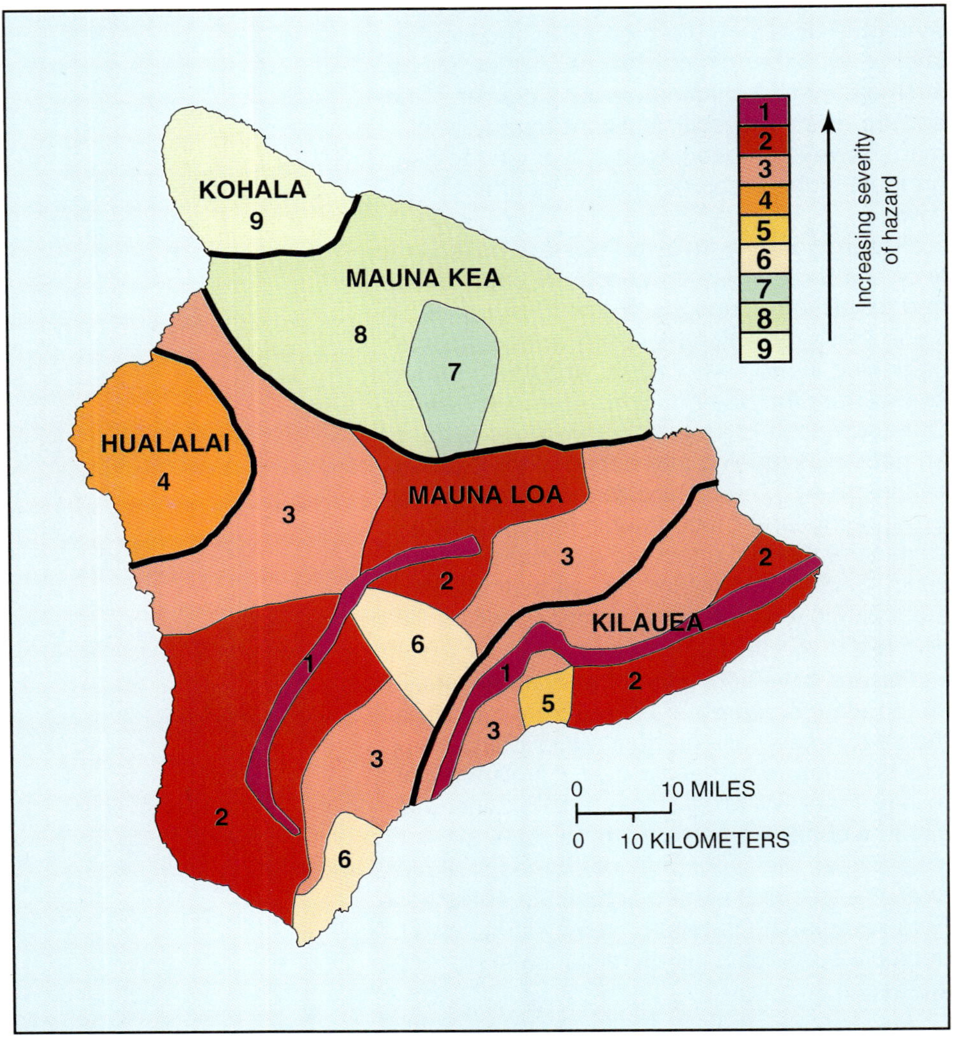

The Island of Hawaii is divided into zones according to the degree of hazards from lava flows: Zone 1 is the area of the greatest hazard, Zone 9 of the least.

than the surrounding area, even though the entire area is included in a single zone. To determine the hazard differences within a single zone, more detailed studies are required.

Hazard Zones for Lava Flows

Zone	Percent of area covered by lava since 1800	Percent of area covered by lava in last 750 years	Explanation
Zone 1	greater than 25%	greater than 65%	Includes the summits and rift zones of Kilauea and Mauna Loa where vents have been repeatedly active in historic time.
Zone 2	15-25%	25-75%	Areas adjacent to and downslope of active rift zones.
Zone 3	1-5%	15-75%	Areas gradationally less hazardous than Zone 2 because of greater distance from recently active vents and/or because the topography makes it less likely that flows will cover these areas.
Zone 4	about 5%	less than 15%	Includes all of Hualalai, where the frequency of eruptions is lower than on Kilauea and Mauna Loa. Flows typically cover large areas.
Zone 5	none	about 50%	Areas currently protected from lava flows by the topography of the volcano.
Zone 6	none	very little	Same as Zone 5.
Zone 7	none	none	20 percent of this area covered by lava in the past 10,000 years.
Zone 8	none	none	Only a few percent of this area covered in the past 10,000 years.
Zone 9	none	none	No eruptions in this area for the past 60,000 years.

Kilauea

Kilauea is Hawaii's youngest volcano and one of the world's most active. Over 90 percent of Kilauea's surface is covered by lava less than 1,100 years old. In historical time, all of Kilauea's eruptions have occurred either in or near its summit caldera, or along the east or southwest rift zones. For the foreseeable future, we can assume that active vents will be limited to these areas.

Since 1955, 28 percent of the area encompassing the east rift zone and the slope south of the rift zone has been covered by lava flows. The latest eruption of the east rift zone began in 1983 and continues as of 1997. The southwest rift zone is less active, with five eruptions in the past 200 years; the latest was in 1974. The most recent summit eruption occurred in 1982.

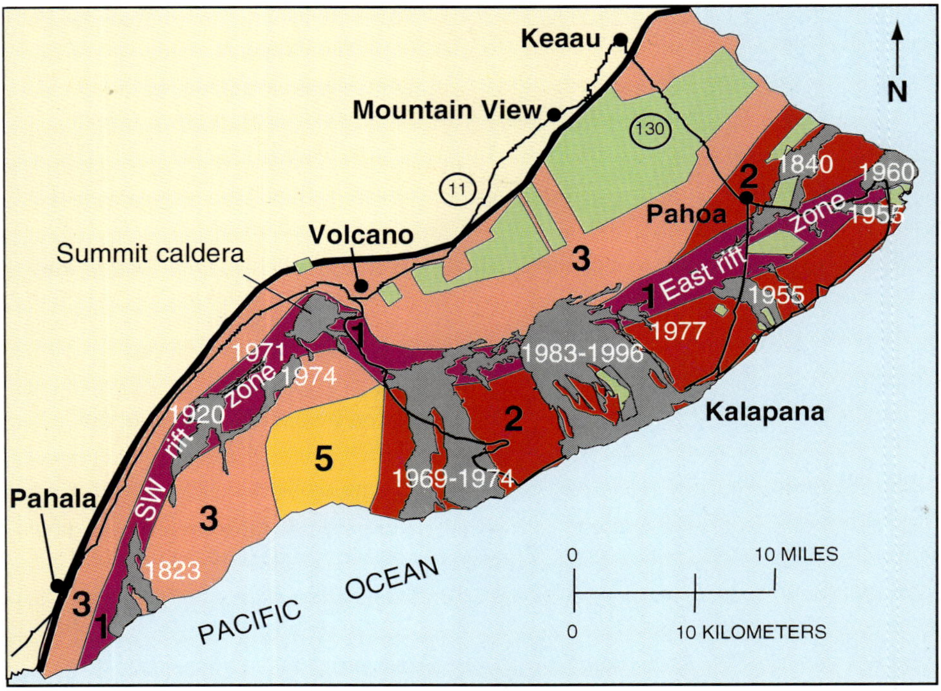

Hazard zones for lava flows on Kilauea. The flows erupted since 1800 are shown in gray and dated. Twenty-eight percent of the area encompassed by Zones 1 and 2 on the east half of the volcano has been covered by lava since 1955. The major housing subdivisions on the slopes of the volcano are shown in green.

From the time of the first written accounts in the early 1800's and through the first decades of this century, Kilauea erupted almost continuously at its summit caldera. In 1924, the summit caldera's active lava lake in Halemaumau crater abruptly drained away, and ground water beneath the caldera apparently came into contact with the hot rocks surrounding the magma conduit, causing a series of steam explosions that threw out large blocks of dense lava along with ash. Sporadic eruptions continued at Halemaumau until 1934, when all activity ceased for 18 years. In 1952, Kilauea reawoke with a 4-month eruption in Halemaumau. Since then, eruptive activity has occurred mainly on the volcano's two rift zones, particularly the east rift zone. From 1969 to 1974, Kilauea erupted at the Mauna Ulu vent on the upper east rift zone. The Mauna Ulu eruption was the most voluminous rift eruption in over a century but has been surpassed in both volume and duration by the recent eruption at the Pu'u 'O'o and Kupaianaha vents.

The hazard map for Kilauea shows the relative degree of hazard from lava flows for different areas of the volcano. Zone 1 is the most hazardous; it consists of the summit area and rift zones because Kilauea's frequent eruptions originate in these areas.

Zone 2 includes the areas that are adjacent to, and downslope from, the east rift zone. The entire area south of the east rift zone lies in this zone. Lava flows are most likely to travel in this direction because the ground slopes downhill from the rift zone to the ocean. The area north of the lower east rift zone, which includes Pahoa, is also in Zone 2. Here the land slopes away to the north as well as the south, and flows can advance in either direction.

Zone 3 includes the areas north of the upper east rift zone and both north and south of the southwest rift zone. Less than 5 percent of the area in Zone 3 has been covered with lava in historical time, but more than 75 percent has been covered in the last 750 years.

No area on Kilauea is ranked Zone 4, but the area south of Kilauea's summit is classed as Zone 5. This flank of Kilauea is currently protected from lava flows by the location of the summit caldera and by north-facing fault scarps (steep slopes or cliffs that form as the result of movement along faults). Because of those topographic barriers, almost none of this area has been affected by lava flows during historical time, although nearly half of it has been covered within the last 750 years.

Steam rises from the Kalapana shoreline where lava flows from Kupaianaha entered the ocean in 1987. The Pu'u 'O'o cone (P) marks the east rift zone; Kupaianaha (K) is the low mound on the center skyline. All of the area from the rift zone to the ocean is in Zone 2. The streets of Royal Gardens are visible on the slope at center.

Mauna Loa

Mauna Loa erupts less frequently than does Kilauea, but it tends to produce a much greater volume of lava over a shorter period of time. Forty percent of Mauna Loa's surface is covered by lava flows less than 1,000 years old. Nearly all of the Mauna Loa eruptions observed since the early 1800's began at its summit caldera; during half of these, the activity subsequently shifted to either the northeast or the southwest rift zone. In addition to the summit and rift zones, the upper northwest flank of Mauna Loa has been the source of three eruptions in the last two centuries. The largest of these, in 1859, produced a lava flow that reached the ocean north of Kiholo Bay on Hawaii's west coast. Mauna Loa has also erupted from a submarine vent on its west flank in historic time. In 1877, a 1-day eruption took place beneath Kealakekua Bay within 1 mile of shore. This eruption produced turbulent water and floating blocks of lava, but it caused no injuries to onlookers who approached the area in canoes and other small boats.

Between 1868 and 1950, lava flows from the southwest rift zone reached the ocean during five eruptions. Flows from four of these eruptions traveled to the sea in 3 to 48 hours. Since 1900, Mauna Loa has erupted 15 times, with eruptions lasting from a few hours to 145 days. After the 1950 eruption, Mauna Loa was quiet for 25 years. It reawakened with a 1-day summit eruption in 1975.

The most recent eruption of Mauna Loa occurred in 1984. This eruption originated at the summit and, within a few hours, migrated to the northeast rift zone. The resulting lava flows advanced to within 4 miles of Hilo before the 3-week-long eruption ended. Similar short-duration eruptions of Mauna Loa's northeast rift zone in 1852 and 1942 produced flows that came within about the same distance of Hilo. In 1855, a much longer-lived eruption fed a flow that stopped half

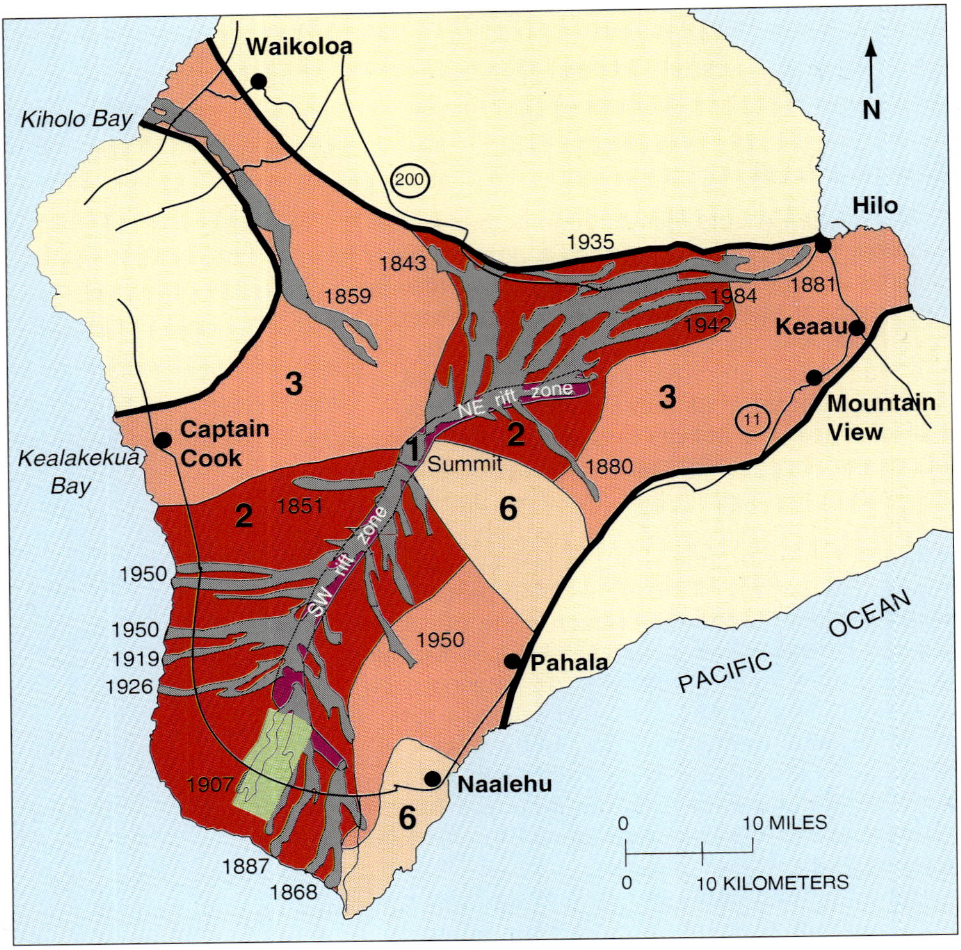

Hazard zones for lava flows on Mauna Loa. The flows erupted in the last 150 years are shown in gray and dated. The major housing subdivisions are shown in green.

a mile east of the upper Kaumana area on the western outskirts of modern Hilo. Lava invaded the present boundaries of Hilo in 1881, although the flow did not reach the shoreline, where the village of Hilo was located at that time. The 1881 flow underlies much of Kaumana and extends a half mile down-slope of Komohana Road.

Zone 1 on the lava flow hazard map for Mauna Loa includes the summit region and the recently active parts of the rift zones.

Zone 2 consists of areas on both sides of the northeast and southwest rift zones. Since both of Mauna Loa's rift zones form prominent ridges, all the areas in Zone 2 are downslope of potential eruption sites. About 20 percent of this area has been covered by lava in historical time, 5 percent since 1950.

Zone 3 includes other areas on Mauna Loa in which the hazard is gradationally lower than in Zone 2. During the past 750 years, lava flows have covered about 15 to 20 percent of Zone 3 on Mauna Loa. These areas are less affected by rift activity than Zone 2, although the area of Zone 3 that lies on the northwest flank of the volcano is vulnerable to eruptions originating at vents on that flank. The 1859 lava flow covers 10 percent of this area. The part of Hilo that lies south of the Wailuku River is included in Zone 3 of Mauna Loa.

Two areas on Mauna Loa are classed as Zone 6 because they are currently protected from lava flows by the local topography. One of these is the Naalehu area; the other is the slope southeast of the present summit caldera.

The 1984 lava flows viewed over the lights of Hilo. The flows advanced to within 4 miles of the town before the eruption ended. *(Photograph by David Little)*

Hualalai

Hualalai is much older than Kilauea and Mauna Loa, and its eruptions occur far less frequently. In the last 3,000 years, Hualalai has erupted near its summit, along the northwest and south-southeast rift zones, and from vents on the north flank of the volcano. Twenty-five percent of the volcano is covered by flows less than 1,000 years old.

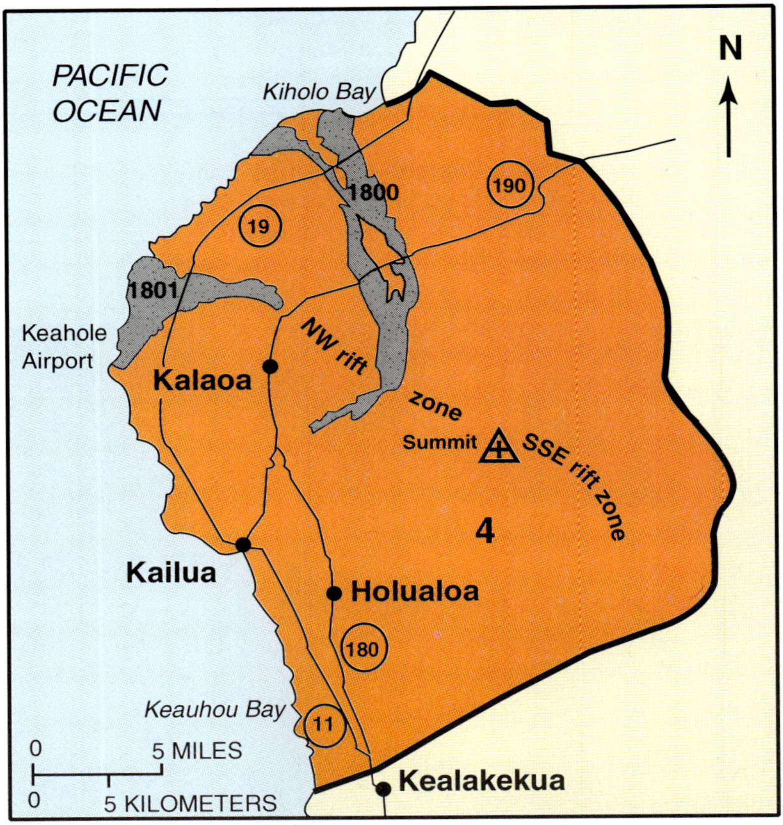

Hazard zone for lava flows on Hualalai. The entire volcano comprises Zone 4. Major flows since 1800 are shown in gray and are dated.

Hualalai last erupted in 1800-1801 from several vents on the northwest rift zone. Large flows spilled down both sides of the ridge formed by the rift zone and quickly reached the ocean. One of these flows lies south of Kiholo Bay, and part of the Kona Village resort is built upon it. Another flow underlies the northern end of the Keahole (Kona) Airport. Other major eruptions occurred about 300 and 700 years ago. A large flow from the 700-year-old eruption forms the north side of Keauhou Bay, south of Kailua.

All of Hualalai is included in Zone 4. The flanks of the volcano do not have a distinctly lower hazard than its rift zones because the distance from the vents to the coast is short and the slopes are steep.

The summit of Hualalai with the town of Kailua in the background.

Mauna Kea and Kohala

The northern part of the Big Island is made up of two volcanoes, Mauna Kea and Kohala. Mauna Kea has erupted several times in the last 10,000 years, most recently about 4,500 years ago. This volcano is considered dormant but not extinct.

The summit and upper slopes of Mauna Kea comprise Zone 7 on the lava flow hazard map. Lava covered about 20 percent of this zone in the last 10,000 years.

Zone 8 includes the lower slopes of Mauna Kea. Most of this area has not been affected by lava flows for the past 10,000 years. The part of Hilo that lies north of the Wailuku river is within this zone.

Zone 9 consists of Kohala, the oldest volcano on the island, which last erupted about 60,000 years ago. Although it is impossible to know if this volcano is extinct or only dormant, the volcanic hazard there is extremely low.

Hazard zones for lava flows on Mauna Kea and Kohala.

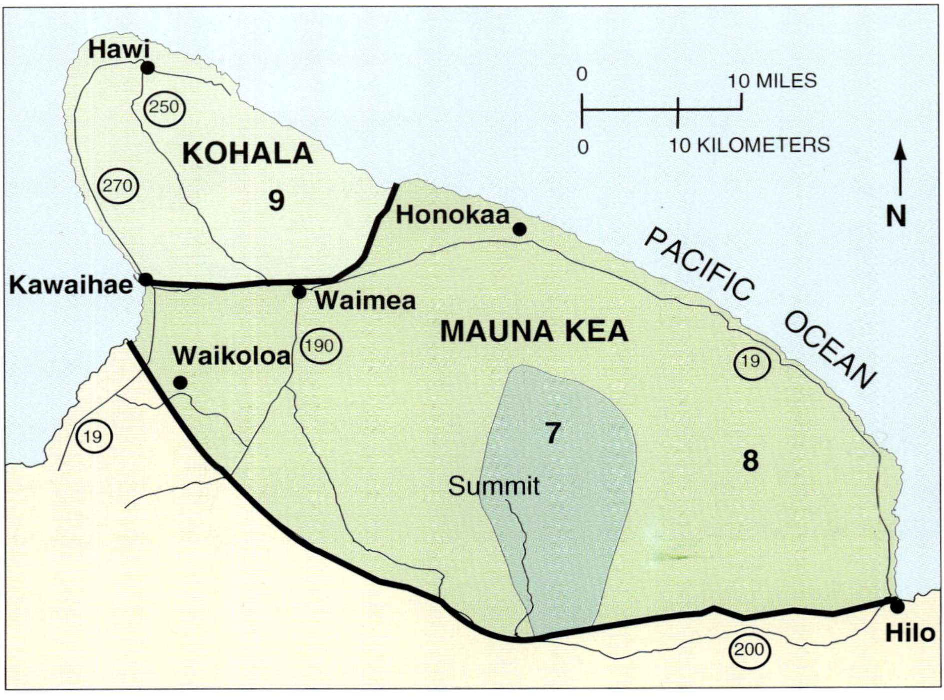

Earthquake Hazards

The Island of Hawaii experiences thousands of earthquakes each year; most are so small that they can only be detected by instruments, but some are strong enough to be felt, and a few cause minor-to-moderate damage. Most of Hawaii's earthquakes are directly related to volcanic activity and are caused by magma moving beneath the earth's surface. Earthquakes may occur before or during an eruption, or they may result from the underground movement of magma that comes close to the surface but does not erupt. A few of

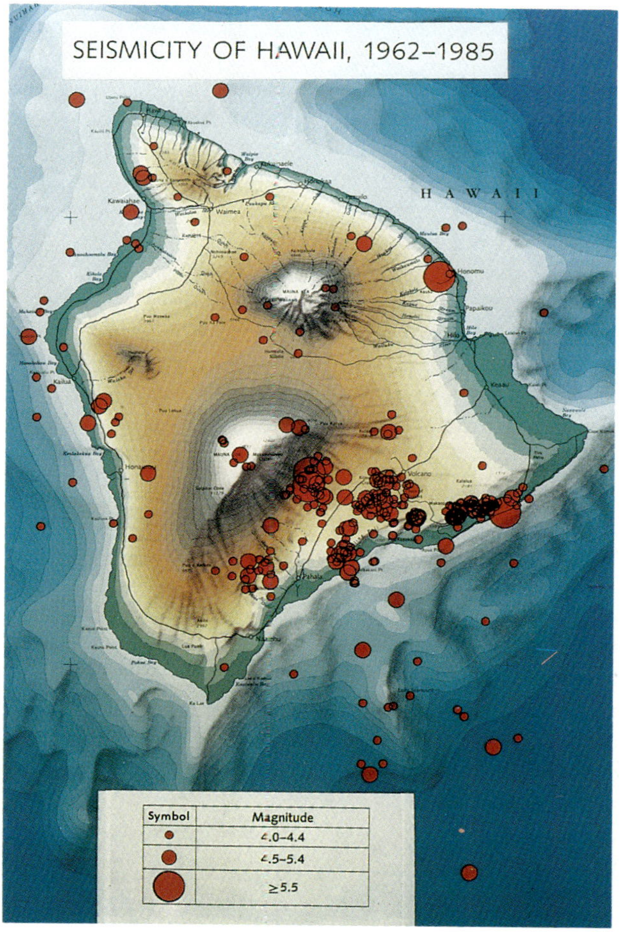

Earthquakes on and near the Island of Hawaii, 1962-1985

the island's earthquakes are less directly related to volcanism; these earthquakes originate in zones of structural weakness at the base of the volcanoes or deep within the earth beneath the island.

Strong earthquakes endanger people and property by shaking structures and by causing ground cracks, ground settling, and landslides. Strong earthquakes in Hawaii's past have destroyed buildings, water tanks, and bridges, and have disrupted water, sewer, and utility lines. Locally, such damage can be intensified where soft, water-saturated soils amplify earthquake ground motions. On steep slopes, such soils may fail during an earthquake, resulting in mudflows or landslides. An indirect hazard produced by some earthquakes is a tsunami, a large sea wave that can be far more damaging than any of the direct seismic hazards.

The size of an earthquake is commonly expressed by its magnitude on the Richter scale, which is a measure of the relative size of the earthquake wave recorded on seismographs. An increase of one whole number on the Richter scale represents a tenfold increase in the amplitude of the seismograph recording. Earthquakes greater than about magnitude 3 usually can be felt by people near the source area, those greater than magnitude 5 are potentially damaging, and any earthquake of magnitude 7 or greater that occurs near populated areas is certain to cause widespread property damage.

The earthquakes directly associated with the movement of magma are concentrated beneath the island's active volcanoes, Kilauea and Mauna Loa. Very shallow earthquakes frequently precede or accompany an eruption. Hundreds of such earthquakes make up swarms that commonly occur over a period of several hours or days before an eruption as magma forces its way into a new area. These earthquakes are seldom large enough to cause widespread damage, but they may produce extensive ground fracturing close to the

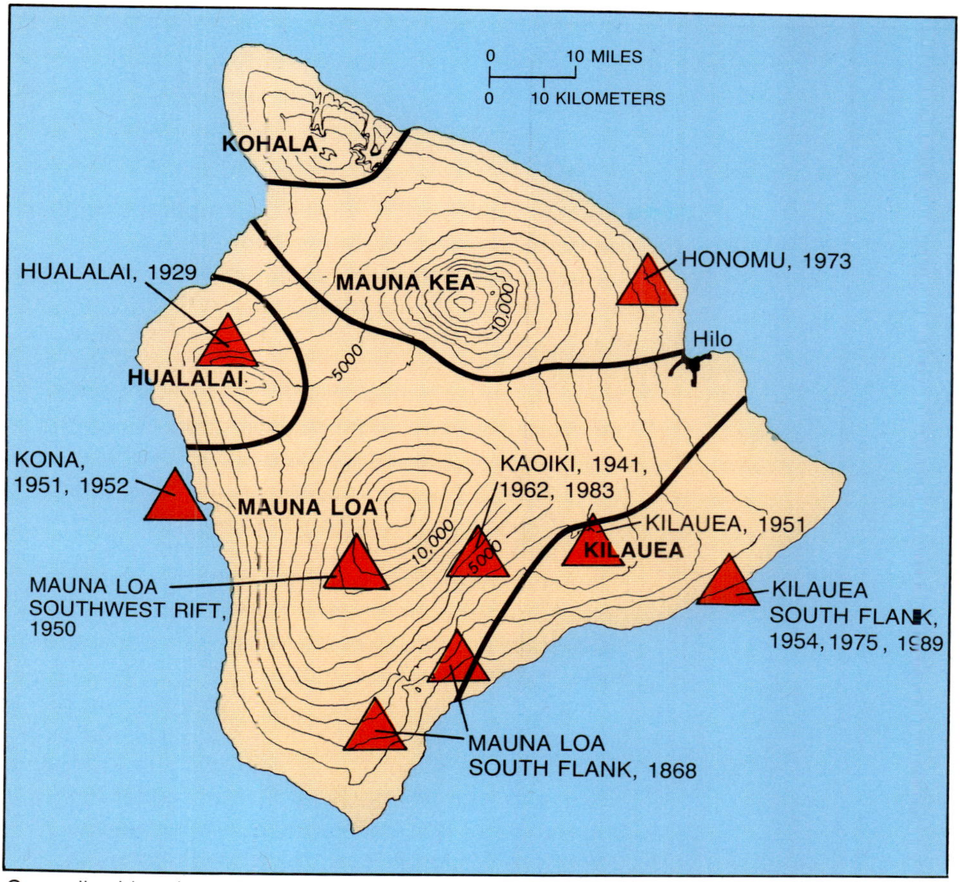

Generalized locations of damaging earthquakes of magnitude 6 or greater since 1868 on the Island of Hawaii. The same earthquakes are listed in the table below. The magnitudes marked by an (*) are based on eyewitness accounts of the earthquakes' effects and on reports of damage. Since about 1950, the accuracy of location, magnitude, and depth determinations has continually improved as the capability of seismic instruments has increased.

Year	Date	Region	Magnitude	Depth (Mi)
1868	Mar. 28	Mauna Loa south flank	6.5-7.0*	No data
1868	Apr. 2	Mauna Loa south flank	7.5-8.1*	No data
1929	Oct. 5	Hualalai	6.5*	No data
1941	Sept. 25	Kaoiki	6.0*	No data
1950	May 29	Mauna Loa southwest rift	6.2	No data
1951	Apr. 22	Kilauea	6.3	20
1951	Aug. 21	Kona	6.9	5
1952	May 23	Kona	6.0	5
1954	Mar. 30	Kilauea south flank	6.5	5
1962	June 27	Kaoiki	6.1	6
1973	Apr. 26	Honomu	6.2	25
1975	Nov. 29	Kilauea south flank	7.2	6
1983	Nov. 16	Kaoiki	6.6	7
1989	June 25	Kilauea south flank	6.1	9

potential eruption site. Once an eruption begins, the earthquakes usually diminish.

Other earthquakes beneath the active volcanoes are generated by the pressures exerted by magma that never reaches the surface. Kilauea's east rift zone is continually being wedged apart by the injection of new magma, much of which is stored underground deep within the rift zone. Since the north flank of Kilauea is immobilized by the adjacent mass of Mauna Loa, the south flank, which faces the ocean, must move outward to make room for the additional magma. Periodically, Kilauea's south flank abruptly shifts seaward in response to this pressure, causing earthquakes. Most such earthquakes are small, but a few are large and can cause damage.

In 1975, Kilauea's south flank was the site of the magnitude 7.2 Kalapana earthquake, the highest magnitude event in this century. The Kalapana coast subsided as much as 11 feet, generating a huge tsunami that claimed two lives in Hawaii Volcanoes National Park, destroyed houses in Punaluu, sank fishing boats in Keauhou Bay, and damaged boats and piers in Hilo. The most recent large earthquake on Kilauea's south flank occurred in June 1989. With a magnitude of 6.1, this quake caused much less damage than the 1975 event.

Earthquakes occur for similar reasons beneath Mauna Loa's southwest and southeast flanks. The Kealakekua fault zone on Hawaii's Kona coast was the site of an earthquake of about magnitude 6.9 in 1951 that may have been related to the 1950 eruption of Mauna Loa's southwest rift zone. The largest Hawaiian earthquake in recorded history occurred in 1868 beneath the Ka'u district on the southeast flank of Mauna Loa; it had an estimated magnitude of between 7.5 and 8.1. The 1868 earthquake caused damage across the entire island and was felt as far away as the Island of Kauai. The devastation was greatest in the Ka'u district, where an earthquake-triggered mudflow killed 31 people and coastal subsidence produced a tsunami that destroyed several villages. At least 79 people perished during this earthquake; most of these casualties resulted from the landslide and tsunami.

Earthquakes in the Kaoiki region, centered between Kilauea and Mauna Loa, are also thought to be related to stresses in the earth's crust that are produced by the activity of the two volcanoes. In the last half century, earthquakes with magnitudes from 5.5 to 6.6 have shaken the Kaoiki region about once every 10 years. The latest large earthquake in this area had a magnitude of 6.6 and occurred in 1983. This event caused substantial damage to structures in Ka'u, Puna, and North and South Hilo districts. Ground cracking and settling led to temporary road closures, and landslides occurred on steep slopes. The financial losses caused by the earthquake were estimated at $7 million. Fortunately, there were only minor injuries

House destroyed in June 25, 1989 earthquake in Kalapana. *(Photograph by J.D. Griggs, USGS)*

because the earthquake struck early in the morning when most people were still in bed.

Large earthquakes unrelated to volcanic activity also occur at irregular intervals on the Island of Hawaii. In 1973, a magnitude 6.2 earthquake located 25 miles beneath Honomu, north of Hilo, injured 11 people and caused $5.6 million worth of damage. Such earthquakes have no known recurrence interval and are difficult to predict.

Defining hazard zones for the effects of earthquakes is more difficult than for eruptions and has not been attempted for the Island of Hawaii. For the most part, earthquakes on Hawaii are concentrated beneath Kilauea and Mauna Loa, particularly beneath the south flanks of both volcanoes, and in the Kaoiki region between them. The likelihood of a damaging earthquake on Kilauea or Mauna Loa probably increases with long-lived activity of the rift zones, but its precise time and magnitude are impossible to predict.

Small, non-damaging earthquakes will be felt more frequently by people living on the slopes of these volcanoes. The effects of a large earthquake under Kilauea or Mauna Loa, however, will not be limited to the immediate area and may cause damage over much of the island. Similarly, deep earthquakes, such as the 1973 Honomu earthquake, can take place under any part of the island and can produce damaging effects over a wide area.

One problem in assigning seismic hazard zones to the island is that the ground shaking during a strong earthquake may vary within a small area. This variation is because the effects of earthquakes are closely related to the nature of the underlying ground; for example, whether it is mainly lava bedrock or soil. Two homes in the same neighborhood may suffer different degrees of damage depending on the properties of the ground upon which they are built. In addition, local topography strongly affects earthquake hazards. Steep slopes composed of loose material may produce large landslides during an earthquake.

The risk from living in a seismically active area, unlike that of living in an area prone to being covered by lava, also depends to a large degree on the type of construction used in a given home. Earthquake shaking may damage certain types of houses, while leaving other types of construction unscathed. For all of these reasons, earthquake hazards are highly localized, and it is difficult to define broad zones with the same relative degree of hazard.

Hilo building damaged by the November 1983 Kaoiki earthquake. *(Photograph by J.M. Buchanan-Banks, USGS)*

Tsunamis

Tsunamis are large, rapidly moving ocean waves triggered by a major disturbance of the ocean floor, which is usually caused by an earthquake but sometimes can be produced by a submarine landslide or a volcanic eruption. Tsunamis are also referred to as "tidal waves," but they have no relation to tides.

Ships at sea cannot detect a passing tsunami, nor can the waves be seen from aircraft. While passing through deep oceans, a tsunami consists of a series of waves that are only a few feet high and a hundred miles or more apart. These waves typically travel at speeds of about 600 mph. As they reach shallow water, the waves slow down but greatly increase in height, and the distance between them shrinks. When the tsunami finally strikes the coast, the waves may crest to heights of 100 feet and travel inland at speeds of 30 mph. A series of waves may reach the coast at intervals of 5 to 40 minutes; the first wave is frequently not the largest.

This map shows how many hours it would take for a tsunami generated anywhere on the rim of the Pacific basin to reach Hawaii. The dots indicate the locations of earthquakes that have generated tsunamis which have affected Hawaii.

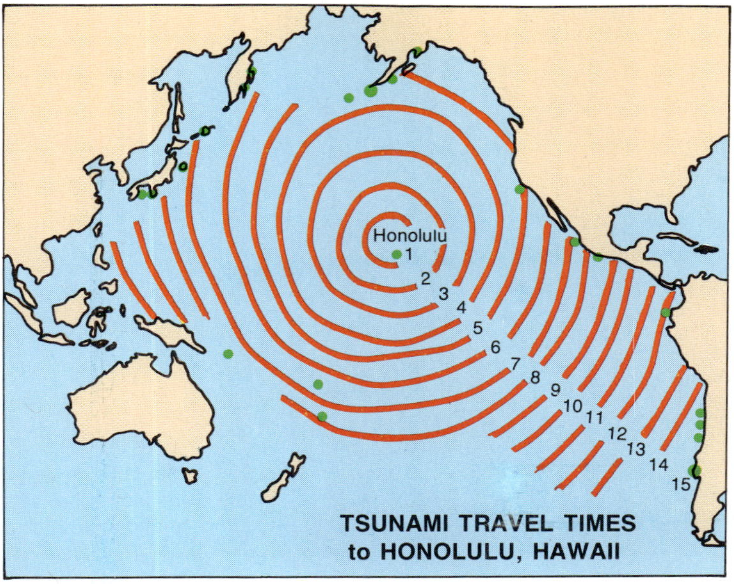

A giant wave engulfs the Hilo pier during the 1946 tsunami. The arrow points to a man who was swept away seconds later. *(Photograph provided by NOAA/EDIS)*

Tsunamis potentially destructive to the Big Island may originate anywhere around the rim of the Pacific Ocean, or they may be locally generated by earthquakes on or near this island. A tsunami produced by an earthquake on the coast of Chile will reach the Hawaiian Islands in about 15 hours, while one that originates in the Aleutian Islands will arrive in 4.5 to 5.5 hours. A locally generated tsunami gives much less warning; the waves may strike almost immediately after the earthquake occurs.

About 50 tsunamis have been reported in the Hawaiian Islands since the early 1800's. Seven caused major damage, and two of these were locally generated. The most devastating tsunamis to hit the Island of Hawaii in this century occurred in 1946 and 1960. In both cases, the worst damage was inflicted on the northeastern coast of the island. The tsunami of 1946 originated in the Aleutian Islands, struck Hawaii without warning, and killed over 170 people, mainly at Laupahoehoe and Hilo where the wave heights averaged 30 feet. The maximum wave height was 55 feet at Pololu Valley on the northern tip of the island.

The 1960 tsunami originated in Chile and advanced upon the island from the southeast; again, its effects were greatest at Hilo. The arrival time of this tsunami was correctly predicted, but many people failed to heed the warnings, and authorities evacuated an insufficient area of Hilo. As a result, 61 lives were lost as waves up to 35 feet high crashed through homes. Whole city blocks were swept clean of all buildings, and 580 acres were flooded.

The tsunamis of 1868 and 1975 were locally generated by earthquakes beneath the southern coast of the island. The 1868 waves destroyed several coastal villages in the Ka'u and Puna districts, most of which were never rebuilt. The 1975 tsunami claimed two lives and caused widespread damage along the Kalapana coast.

Following the disastrous tsunami of 1946, a tsunami warning system for the Pacific basin was developed by the U.S. Coast and Geodetic Survey. Presently, the Pacific Tsunami Warning System, which has its headquarters in Honolulu, is administered by the National Weather Service under the National Oceanic and Atmospheric Administration (NOAA). The Tsunami Warning Center receives a constant flow of information from seismometers (instruments that detect earth-

Tsunami destruction on Kamehameha Avenue on Hilo's waterfront, 1946. *(Photograph provided by the U.S. Army Corps of Engineers)*

quakes) and tide-gauge stations all around the Pacific Ocean. When an earthquake with the potential to generate a tsunami occurs, the Warning Center puts out a "tsunami watch" within 30-40 minutes, alerting Civil Defense and other authorities of a possible tsunami. The first positive evidence of a tsunami comes from tide stations near the origin of the earthquake. Once the presence of a wave is confirmed, the Honolulu center issues a "tsunami warning," giving an estimated time of arrival for the first wave. When a warning is issued, local authorities evacuate low-lying coastal areas.

Because the speed of a tsunami depends entirely on the depth of the water, the arrival time of a wave from any point on the Pacific rim can be predicted. The size of the wave is difficult to predict, however, because the Hawaiian Islands may be the first place to be hit by the wave. The effects of each new tsunami, and the effects on different coasts of the same tsunami, vary greatly. The size and destructiveness of the waves at any particular site are largely determined by the local topography, both onshore and offshore, and the direction from which the wave approaches. For these reasons, people should not become complacent just because one tsunami warning ends with only a few inches' rise in water level. The next time, a warning could precede a 55-foot wave.

Locally generated tsunamis, such as those of 1868 and 1975, are potentially the most hazardous type, because the time between their origin and the arrival of the wave at the shoreline may be too brief to warn and evacuate people. In 1975, the first wave reached Punalu'u immediately after the earthquake; it arrived at Hilo in 20 minutes. Any earthquake strong enough to cause difficulty in standing or walking should be regarded as a tsunami warning by people in coastal areas, who should immediately head for higher ground.

Living with Volcanic and Seismic Hazards

The best defense against any natural hazard is education. Residents and public officials should be aware of the hazards in a given area so that they can make rational decisions regarding where to build homes, develop property for commercial use, and locate public facilities. A well-informed public will neither over-react to the hazards nor ignore them.

When an area is threatened by a nearby eruption or an approaching lava flow, the hazard to life can be minimized by evacuating people. The repeated evacuation of residential areas near active rift zones comes at a high cost, however, both for the county government and taxpayers and for the people whose property is threatened by lava. For the long term, volcanic hazards can be avoided by land-use zoning that restricts building in the areas of highest hazard.

The Mauna Loa eruption of 1984 fed a major lava flow that advanced toward the city of Hilo. *(Photograph provided by National Park Service)*

The Wahaula Visitor Center in Hawaii Volcanoes National Park was overrun by lava flows in June 1989. *(Photograph by J.D. Griggs, USGS)*

Diverting lava flows by artificial means is a largely untested and costly option for protecting developed areas. Well-placed barriers may successfully divert a short-lived lava flow, but during a longer eruption, keeping up with the sheer volume of lava and the number of flows involved may prove impossible. During the 1955 eruption of Kilauea's east rift zone, barriers temporarily diverted flows from two different plantations, but in both cases, flows on subsequent days took different routes and ultimately destroyed the property.

Although a few attempts at lava diversion have been partially successful elsewhere in the world, such efforts require favorable conditions of topography and property ownership that can rarely be met in a populated area. Artificial diversion of lava onto property that otherwise would have been spared could lead to complex legal problems. Lava diversion, however, is a reasonable option in unpopulated areas where isolated, high-value property is at risk. For example, diversion structures have been constructed in Hawaii to protect the Mauna Loa Observatory, NOAA's atmospheric research station, from future lava flows.

The potential for damage from strong earthquakes cannot be avoided on the Island of

Hawaii. The damage could be reduced, however, by land-use zoning that restricts building on or near steep slopes that may fail during earthquakes and in areas underlain by materials that are likely to amplify the ground motion of a strong earthquake. The Island of Hawaii is in Zone 4 of the Uniform Building Code, which requires public and certain types of private buildings to meet structural design standards for earthquake resistance. Information about the Building Code can be obtained from the Hawaii County Building Division of the Public Works Department.

The magnitude 6.6 Kaoiki earthquake in 1983 caused ground cracks and landslides on the rim of Kilauea's summit caldera. *(Photograph by J.D. Griggs, USGS)*

Tsunamis are a potential threat to all low-lying coastal areas of the island. The tsunami risk to new developments can be minimized by restrictive zoning, such as already exists on Hilo's waterfront, and by flood-resistant construction. Since much of Hawaii's coast is already developed, evacuation of low-lying areas in response to tsunami warnings is the only practical option for avoiding loss of life.

More information about the tsunami warning system, evacuation routes, and emergency procedures can be obtained from the Hawaii County Civil Defense.

Tsunami destruction in Hilc, May 1960. *(Photographer unknown)*

Selected Readings

Decker, R.W., Wright, T.L., and Stauffer, P.H., 1987, editors, Volcanism in Hawaii: U.S. Geological Survey Professional Paper 1350, vol. 1. *(This volume contains the following four papers, from which much of the information in this publication is derived.)*

Holcomb, R.T., 1987, Eruptive history and long-term behavior of Kilauea Volcano, p. 261-350.

Lockwood, J.P., and Lipman, P.W., 1987, Holocene eruptive history of Mauna Loa Volcano, p. 509-535.

Moore, R.B., Clague, D.A., Meyer, R., and Bohrson, W.A.,1987, Hualalai Volcano: A preliminary summary of geologic, petrologic, and geophysical data, p. 571-585.

Mullineaux, D. R. , Peterson, D.W., and Crandell, D.R.,1987, Volcanic hazards in the Hawaiian Islands, p. 599-621.

Macdonald, G.A., Abbott, A.T., and Peterson, F.L., 1983, Volcanoes in the sea: The geology of Hawaii: University of Hawaii Press, Honolulu, 517 p.

Tilling, R.I., Heliker, C., and Wright, T.L., 1987, Eruptions of Hawaiian volcanoes: Past, present, and future: U.S. Geological Survey general interest publication, 54 p.

Wright, T.L., Chun, J.Y.F., Esposo, J., Heliker, C., Hodge, J., Lockwood, J.P., and Vogt, S., 1992, Map showing lava flow hazard zones, Island of Hawaii, 1:250,000: U.S.G.S. Misc. Field studies map MF-2193.

U.S. Geological Survey
Information Services
P.O. Box 25286
Denver, CO 80225

Inside Back Cover: Lava enters the ocean on Kilauea's southern coast in October 1988.

Back Cover: *(Photograph by J. D. Griggs, USGS)*

First printing, 1990; revised, 1992; revised, 1997.

☆ U.S. GOVERNMENT PRINTING OFFICE: 1997 426-748